THE PLANET-FRIENDLY KITCHEN

THE PLANET-FRIENDLY KITCHEN

An Hachette UK Company
www.hachette.co.uk

Summersdale Publishers Ltd
Part of Octopus Publishing Group Limited
Carmelite House
50 Victoria Embankment
LONDON
EC4Y 0DZ
UK

www.summersdale.com

Printed and bound in China

ISBN: 978-1-78783-691-4

Substantial discounts on bulk quantities of Summersdale books are available to corporations, professional associations and other organizations. For details contact general enquiries: telephone: +44 (0) 1243 771107 or email: enquiries@summersdale.com.

Statistics are all correct as of September 2020

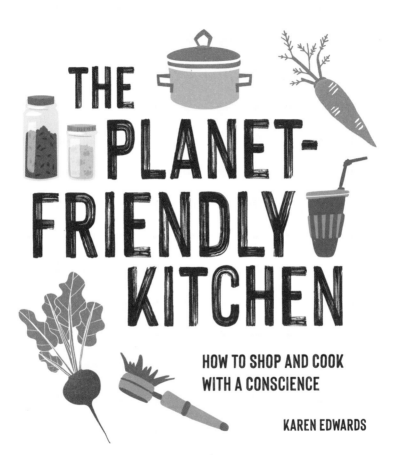

THE PLANET-FRIENDLY KITCHEN

HOW TO SHOP AND COOK WITH A CONSCIENCE

KAREN EDWARDS

summersdale

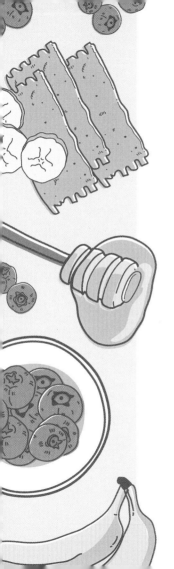

CONTENTS

Introduction — 6

Part 1: What Is Planet-Friendly Food? — 8

Where does our food come from? — 14

The responsible consumer — 34

The environmentally friendly kitchen — 50

Part 2: Planet-Friendly Recipes — 60

Conclusion — 126

INTRODUCTION

For many of us, the desire to be more planet-friendly comes from the understanding that the earth's natural resources are finite. Over time, as our collective footprint becomes deeper, we know the land, water and energy required to maintain human life, particularly through food growth and production, cannot always be renewed or replenished.

With the global population expected to rise to more than 9 billion people by 2050, our food habits must be considered if we are to take pressure off the environment. By adapting to a more environmentally conscious lifestyle, we would be making a huge step toward a sustainable future. This includes extending our compassion to the animals that suffer as a result of our diets, from the livestock that are reared, transported and slaughtered for meat consumption to the overfished and endangered marine fauna – not forgetting the terrestrial wildlife left without homes when their habitats are cleared for agricultural land.

This book will shed light on food growth and production, highlighting the produce that requires fewer natural resources to grow and advising on which items are best to avoid. While there is no single definition of "sustainable food", our goal is to understand the concept of a planet-friendly diet by looking at what happens behind the scenes during the food production process. The book will also discuss how to grow your own produce and share many environmentally conscious recipes.

Armed with this knowledge, you will be empowered to make positive, planet-friendly choices as a consumer – remembering that, while most options aren't 100 per cent sustainable, the production of some foods is better for the planet than others. By incorporating these items into our diet and becoming more mindful in our selections, we will be helping the planet thrive for generations to come.

Part 1

WHAT IS PLANET-FRIENDLY FOOD?

PLANET-FRIENDLY FOOD IS...

- Locally grown produce that requires less energy for storage and transport
- Naturally high-yielding crops
- Produce that requires minimal water
- Food that needs minimal processing
- "Nitrogen-fixing" crops, which capture atmospheric nitrogen and return it as nutrients to the soil
- Meat from organic farms, which provide kinder conditions for livestock
- Seafood that is not from an endangered or threatened population
- Produce grown on established farms to avoid further land clearance
- Seafood from abundant stocks over multiple locations
- Products in biodegradable or reusable packaging
- Food that isn't genetically modified or given growth hormones
- Plants that don't require synthetic fertilizers or pesticides
- Crops grown in varied climates, reducing pressure on any one region
- Fair-trade products, certifying that farm workers are treated fairly

Becoming a planet-friendly consumer means knowing exactly where our food comes from, the ingredients it contains and how it has been produced.

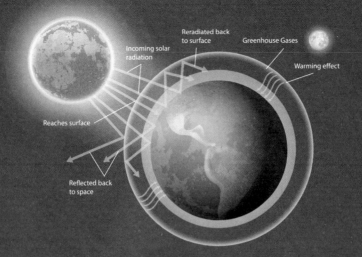

Incoming solar radiation

Reradiated back to surface

Greenhouse Gases

Warming effect

Reaches surface

Reflected back to space

THE IMPACT OF FOOD PRODUCTION ON THE CLIMATE

Most of our food is processed in some way before reaching our tables. This production line – from the farm to the kitchen – generates "greenhouse gases", such as carbon dioxide (CO_2), methane (CH_4) and nitrous oxide (N_2O). Farm machinery, factory processing, refrigeration and transport use energy from the burning of fossil fuels, such as coal and diesel, and produce large amounts of carbon dioxide. Methane is emitted through the fermentation of food in the stomachs of livestock, while nitrous oxide is created through both the burning of fossil fuels and the breakdown of nitrogen-based fertilizers used on crops. The gases absorb radiation from the earth's surface, trapping heat within our atmosphere. This contributes to our warming climate.

It is estimated that between 20 and 30 per cent of greenhouse gases emitted globally through human activity comes from food production.[*]

*Source: Sustain: The Alliance for Better Food and Farming.

THE IMPACT OF FOOD PRODUCTION ON THE ENVIRONMENT

Transportation emissions

Our food comes from all over the world, from the wild fisheries of Alaska to the fruit plantations of Fiji. Cargo planes, ships, trains and trucks delivering food internationally allow us the privilege of enjoying produce previously only available in specific regions or climates. Unfortunately, the burning of fuel during transportation releases high levels of CO_2 into the atmosphere.

Single-use packaging

Plastic pollution is one of the world's most disturbing environmental problems. Due to its durable nature, plastic can remain in the environment for centuries. It is estimated that over 8 million tonnes of plastic reach our oceans each year[*], making packaging an overwhelming danger to marine life.

Deforestation

An area of previously undisturbed forest the size of one football pitch was lost every 6 seconds in 2019, leaving wildlife without homes and fewer trees to absorb CO_2 from the atmosphere. Agriculture is a major cause of deforestation. Today, just 30 per cent of the earth's land area is covered in trees[**], compared to an estimated 50 per cent coverage in the pre-industrial age.

[*]Source: Dr Jenna R. Jambeck, University of Georgia, *Science* journal (2015).

[**]Source: Global Forest Watch, Food and Agriculture Organization of the United Nations.

Water: the foundation of food production

Everything needs water to grow. Irrigation, or the controlled watering of plant crops, is used in environments where rainfall is insufficient to sustain growth. This allows enormous volumes of food to be produced globally. In fact, 40 per cent of the world's food relies on irrigation.[*]

Water used in farming is generally sourced from rivers, lakes and aquifers. Raising livestock demands a greater supply of water than growing crops. While irrigation gives us great scope for food production, the process can significantly affect an ecosystem. For example, excessive water removal can cause nearby wetlands to dry up, leading to wildlife habitat loss.

How much water produces one kilogram of...?

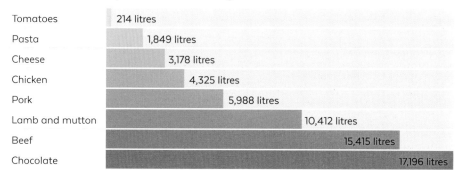

Tomatoes	214 litres
Pasta	1,849 litres
Cheese	3,178 litres
Chicken	4,325 litres
Pork	5,988 litres
Lamb and mutton	10,412 litres
Beef	15,415 litres
Chocolate	17,196 litres

Source: Institution of Mechanical Engineers, "Global Food: Waste Not, Want Not" (November 2013).

[*]Source: UNESCO.

WHERE DOES OUR FOOD COME FROM?

Around 50 per cent of the planet's habitable land is currently used for plant and livestock agriculture.

Food and Agriculture Organization of the United Nations (2019)

FEEDING THE WORLD

Humans have been growing cereal and root crops for around 11,500 years. Livestock farming has been practised for almost as long. Today, some farmers in developing countries still carry out plant cultivation and animal farming just as their ancestors did, usually without machines or chemicals.

Commercial agriculture, on the other hand, requires energy and water-intensive methods for mass production. Soil degradation and the eutrophication (nutrient pollution) of waterways are also associated with large-scale farming. As the global population increases exponentially, the agricultural sector faces an unprecedented demand on produce, putting more pressure on the planet's natural resources.

MEETING DEMAND AT MINIMAL COST

Intensive farming is the practice of producing a maximum yield on a large scale, at a minimal financial cost and while using limited land space. This involves:

- Factory farming or raising high densities of livestock in a confined space
- Increased use of fertilizers and pesticides to encourage flawless crop growth
- Use of genetically modified, high-yielding seeds for crops
- Feeding hormones to animals to create bulkier livestock
- Wide-scale irrigation, requiring a large natural water supply

Is it planet-friendly?

Intensive farming, particularly of crops, could reduce the need for further land clearance and deforestation, while high yields mean global food shortages are potentially prevented. Therefore, it could be considered a sustainable farming method. However, with more livestock suffering in deplorable conditions, a greater use of heavy machinery and the excessive use of fertilizers and pesticides causing heavy pollution of natural waterways, the practice is currently far from planet-friendly.

THE WORLD'S MEAT MARKET

The global demand for pork, beef, chicken and lamb has put an enormous – potentially irreversible – strain on natural resources. As well as using huge amounts of water and energy, animal agriculture has meant large-scale land clearance of the earth's most important ecosystems, including the Amazon rainforest, North America's Great Plains, southern African savannahs, and Australia's subtropical and temperate bushland. This land is used to rear livestock *and* grow enough crops – usually soya and grain – to feed them. As demand for meat grows, vital land and ecosystems will continue to be cleared.

Animal farming alone uses 77 per cent of the world's agricultural land and produces around 55 per cent of agricultural emissions of greenhouse gases. However, it only supplies us with one fifth of the calories we consume.*

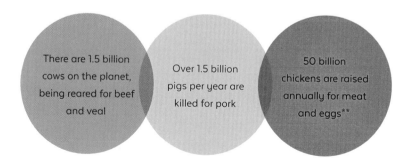

There are 1.5 billion cows on the planet, being reared for beef and veal

Over 1.5 billion pigs per year are killed for pork

50 billion chickens are raised annually for meat and eggs**

*Sources: Food and Agriculture Organization of the United Nations and the United States Environmental Protection Agency.

**Source: World Economic Forum.

LIVESTOCK – MORE THAN JUST MEAT

Unfortunately, animals reared for meat and dairy are habitually bred in questionable conditions. Overstocking is the most common issue. Often animals do not have enough access to food or space to behave naturally. Some livestock are held in pens without natural light; others are given hormones to promote growth. Overstocking also allows disease to spread easily, which is terrible for the animals, but also increases the chances of human infection and transmission.

It's not fun being a veal calf

Male calves of dairy cows are separated from their mothers soon after birth and transferred to veal farms. Here, they are fed milk rather than solid food to ensure their meat remains pale and tender. This lack of iron intake can result in anaemia. On intensive farms, veal calves are sometimes kept in narrow "veal crates", where they are tied by the neck and unable to turn. These conditions cause a great deal of distress. While some regions around the world have banned the cruel practice, it is still a widespread farming method.

Pigs and their pens

Most pigs are raised on factory farms where females are kept in cramped crates and often forced to breed continuously for up to four years. Without the ability to move freely, they are left exhausted, developing muscular and cardiovascular ailments. Reports of neglect are common. In extreme cases, starving pigs have been left to feed on each other.

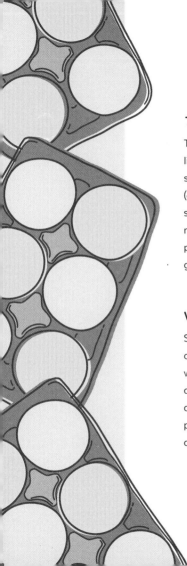

The free-range egg debate

The "free range" label doesn't guarantee a chicken quality of life; it merely confirms the producer has provided an outdoor space, which the bird may not be able to access regularly (or at all). Females raised for egg-bearing are generally still slaughtered when they no longer produce eggs, while males are slaughtered at birth. Organic eggs are likely to be produced using kinder farming methods, although there are no guarantees.

What can we do?

Supporting organic farming, or local independent farms where you can see for yourself how animals are raised, encourages livestock welfare – however, these practices are not currently viable to feed a global population. Teaching ourselves to rely predominantly on a plant-based diet would be the most effective way of reducing pressure on natural resources *and* reducing the suffering of animals reared for meat.

SEAFOOD – DOES IT ALL COME FROM THE SEA?

Due to demand, the wild stocks of many seafood species are diminishing worldwide. To take the pressure off wild fisheries, aquaculture – where seafood is farmed and harvested in a controlled environment – has become an increasingly popular industry. Some aquaculture farms are established in marine environments, while others are created on land but seek to replicate the conditions of a wild aquatic setting. Aquaculture currently produces about half of all seafood consumed around the world.[*]

Which is best – farmed or wild?

Neither method has proven to be more planet-friendly than the other. Both sources have positive and negative effects on the planet. Our main duty as consumers is to understand these effects and ensure that we buy from a local, reliable and sustainable source.

Check these apps for sustainable, preferably local, fisheries before you buy:

Marine Conservation Society Good Fish Guide

Monterey Bay Aquarium Seafood Watch

[*]Source: National Oceanic and Atmospheric Administration (2020).

AQUACULTURE – THE FARMING OF FISH

The positives

- The rearing and harvesting of large numbers of seafood species at a fast rate caters for the growing human population without depleting ecosystems
- Harvesting from an enclosed environment is less stressful for fish
- Popular species can be grown far from their natural environment to save long-distance transportation (for example, Atlantic salmon in Australian waters)
- Takes the pressure off wild stocks

The negatives

- Disease and parasites from fish kept in sea enclosures can be transferred to wild fish
- Excess nutrients, chemicals and antibiotics in fish feed and waste can pollute water systems
- Escaped farmed fish may interbreed with wild stocks, compromising genetic diversity
- The overharvesting of bait fish for farm feed can deplete natural stocks and affect ecosystems

Thumbs up for farmed shellfish

Recent studies suggest that oysters, mussels and clams cause minimal environmental impact when farmed because they do not require wild stock feed or create waste.* In fact, by filter-feeding as normal, they keep the water clean.

*Source: National Oceanic and Atmospheric Administration.

GOING WILD?

If you choose to buy from wild fisheries:

- Avoid threatened or endangered species, including Chilean sea bass, marlin and bluefin tuna
- Buy directly from local markets so the source is guaranteed
- Pick domestic species to reduce demand on overfished international fisheries
- Choose seafood that has an abundant population and a sustainable reproduction rate
- Buy from fisheries using practices that reduce habitat impact and by-catch. Methods such as pole or hand-line fishing are less damaging and cruel than trawl nets
- Don't buy large predatory fish, such as marlin, shark or mackerel, that can contain high compounded levels of mercury
- Avoid caviar! Collecting fish eggs means catching fish before spawning, slicing the roe sack and extracting eggs – which kills the fish. The beluga sturgeon is critically endangered thanks to humanity's fondness for caviar

Say no to krill oil!

The overharvesting of krill (tiny crustaceans) could lead to the demise of entire ecosystems. Antarctic and sub-Antarctic species – such as penguins, whales and seals – depend on krill for their survival.

THE GOOD FISH GUIDE:
WHAT YOU NEED TO KNOW ABOUT POPULAR FISH

Salmon

- There are seven species of Pacific salmon, and one species of Atlantic salmon
- Wild Pacific salmon is currently fished sustainably
- Avoid wild Atlantic salmon stocks that are heavily depleted. Farmed Arctic char is a sustainable Atlantic alternative
- Pacific salmon is best, but consider that these fisheries are coming under greater pressure

Halibut

- Wild-caught Pacific halibut from the north Pacific and British Columbia is sustainable
- Atlantic halibut is sustainably farmed in the UK
- Wild-caught Atlantic halibut stocks, including those from Greenland and the north-east Arctic, are heavily overfished – avoid these

Cod

- There are healthy populations throughout the Atlantic, but some stocks are poorly managed
- Avoid cod taken from depleted stocks – look for the blue MSC label from the Marine Stewardship Council

Tuna

- Albacore tuna is the most sustainable species; skipjack is sustainable if caught using pole and line, or troll methods
- Atlantic yellowfin populations are currently viable; Indian Ocean stocks are heavily overfished – check for the MSC label
- Avoid the currently endangered bluefin

WHAT ABOUT DAIRY?

Around 270 million cows are reared annually for dairy farming across the world.* One cow can produce up to 8,000 litres of milk per year.** To reach such targets, she is expected to almost constantly lactate from the age of two years, after giving birth to her first calf. The calf is taken away within 36 hours so her milk can be collected. She is then artificially inseminated, two to three months post-birth, so the cycle can continue.

The reality of dairy farming

- Artificial impregnation and perpetual lactation cause the cow distress, exhaustion and infections such as mastitis
- Cows naturally live for around 20–22 years but dairy cows are slaughtered at around five years old, when they stop producing milk
- Male calves that aren't sidelined to produce veal are slaughtered at birth

*Source: World Wildlife Fund.

**Source: Agriculture and Horticulture Development Board UK.

What about goat's or sheep's milk?

Often recommended as an alternative milk for those who struggle to digest cow's milk, goat's and sheep's milk contain less of the protein casein, which is known to irritate the digestive system. Unfortunately, the milder effect on our gut doesn't help the goats and sheep that are farmed for their milk in the same unfavourable way as dairy cows.

Do we need dairy in our diets?

Dairy is a convenient source of calcium, potassium and vitamin D. However, we can get the same nutrients from other food sources and supplements. Dairy products contain calcium because the animals have absorbed it from plants. By consuming veggies, we are going straight to the source. Most supermarkets and restaurants offer dairy-free substitutes. In fact, it has never been easier to follow a dairy-free diet.

SOY MISUNDERSTOOD

The United States, Argentina and Brazil combined produce around 80 per cent of the world's soya. It is found in 27 per cent of our vegetable oils and is increasingly used in biofuel production. However, it is a common misconception that soya is mostly consumed by vegetarians and vegans, when in reality 75 per cent of soya farmed around the world is grown for animal feed rather than human consumption. That's a whole lot of soya!

Pigs raised for pork, for example, are typically fed soya beans rather than grass, which creates an excessive demand for soya plantations. As the global population grows and demand for meat increases, more land will be cleared. This is another example of how cutting down our meat consumption and switching to a predominantly plant-based diet could help to reduce pressure on the planet.

Many scientists suggest that a predominantly plant-based diet for humanity could reduce land use by nearly three quarters.

Soya farming is the second largest cause of tropical deforestation, after beef farming.*

*Source: World Wildlife Fund.

PLANT MILK – WHICH IS BEST?

Producing plant milk requires less land, emits fewer greenhouse gases and needs less water than producing animal milk, making *all* plant milks more sustainable than dairy. Plant-based alternatives are also healthier for us, as they contain no cholesterol or animal bacteria. While there is not enough evidence to show conclusively which plant milk is the most sustainable overall, we can identify which milks are best to buy depending on *where* we are consuming them.

Oat, hemp and flax milk are currently responsibly produced in the northern hemisphere, while rice milk and certified sustainable coconut milk are preferable in the southern hemisphere. Soya milk is also highly sustainable in most climates. Almond milk is the least planet-friendly non-dairy option, as 80 per cent of almonds are grown in the arid Californian landscape, requiring a particularly large amount of water for irrigation to enable mass-scale production. The exception is in Australia, where almonds grown locally need much less irrigation.

The environmental impact of one glass of milk

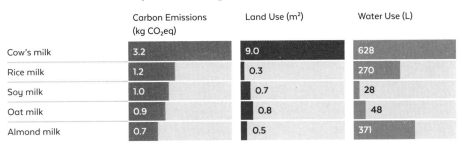

	Carbon Emissions (kg CO_2eq)	Land Use (m^2)	Water Use (L)
Cow's milk	3.2	9.0	628
Rice milk	1.2	0.3	270
Soy milk	1.0	0.7	28
Oat milk	0.9	0.8	48
Almond milk	0.7	0.5	371

It takes 28 litres of water to produce one litre of soya milk and 371 litres of water to make one litre of almond milk.*

*Source: Poore & Nemecek, "Reducing food's environmental impacts through producers and consumers", *Science* (June 2018).

PLAYING WITH GENETICS

What are GMOs?

Genetically modified organisms (GMOs) are living things – such as plant crops and livestock – the DNA of which has been altered using genetic engineering processes.

Why is our food genetically modified?

Organisms are genetically modified to produce a higher yield, give a longer lifespan or offer a higher nutritional value. The most common modification, however, is to introduce resistance in crops and livestock against diseases caused by insects and viruses. This means there is less crop wastage and that fewer pesticides are needed, which is admittedly beneficial for the planet. Genetically modifying livestock promotes faster growth, increased milk production and bulked-up meat. As a result, GM foods are highly profitable for producers.

Are GM foods safe?

The risks around modifying livestock genetics are currently unknown. However, altering the natural genetics in any animal remains controversial. Food standards, including the sale of GM foods, vary by region, and until we have more scientific evidence, consumption is a personal choice. Look for "GM-free" food labels if you wish to avoid GMO products.

IS ORGANIC REALLY BEST?

Organic foods are farmed products that are grown or reared without the use of man-made fertilizers, pesticides, additives or growth hormones. In the case of livestock, antibiotic use is prohibited, and even feed cannot be synthetically fertilized or genetically modified.

The pros and cons of organic farming

Pros

- Free from harmful chemicals and pesticides
- No food additives or hormones
- Not genetically modified
- Livestock are fed a natural diet and kept in better, less cruel conditions
- Generally higher nutritional value

Cons

- Without chemical use, yields are much lower and therefore less effective in feeding large populations
- Shorter shelf life, possibly resulting in waste
- Lower yield means consumer costs remain higher
- Crops and livestock are more prone to disease
- Costly farming methods mean higher prices for the consumer

Unfortunately, organic food is not always sustainable. However, it can be considered planet-friendly. The major benefit is that products are free from synthetic substances, meaning the production process is kinder to the environment and livestock and the food is healthier for human consumption.

A MATTER OF CHOICE

There's no doubt that agriculture is vital to sustain the growing world population. It allows food to be produced on a large scale and is crucial to the livelihoods of around 1 billion people. However, we cannot ignore the significant negative impact that agriculture – and particularly livestock farming – has on the welfare of our planet.

In this section, we have learned that collectively retraining our nutritional preferences around a plant-based diet could save natural resources, ecosystems and wildlife – not to mention giving livestock a kinder existence. While we don't have to switch to a wholly meat-free diet, being more conscious in sourcing meat, and eating less of it, is important.

Supporting local farmers is key, especially those who acknowledge and adopt sustainable, organic methods of farming – because it encourages environmental awareness and ethical practices. This will also urge governments to see the value in planet-conscious farming going forward.

IS GOING VEGAN THE ANSWER?

Eating less animal-derived produce reduces our impact on the planet because it allows us to:

- Save the lives of wildlife whose habitats are lost through land clearance
- Decrease the numbers of livestock reared for dairy or meat
- Lower carbon emissions that contribute to climate change
- Reduce deforestation and land degradation dramatically
- Lessen the overall demand on natural resources

Making a difference

Adopting a vegan diet is the most effective step toward a more planet-friendly food industry. Even simply choosing to eat plant-based meals twice a day (or 730 animal-free meals in a year) can make a significant difference to the demand on meat production. If most of the population did this, the reliance on livestock agriculture would decrease dramatically.

THE HEALTH BENEFITS OF
A PLANT-BASED DIET

It's not all about being more planet-friendly. Eating more plant-based food is also better for our health. A study funded by the British Heart Foundation found that a balanced plant-based diet can reduce premature mortality by 18 to 24 per cent.* Other potential benefits are:

- Better concentration, more energy, reduced irritability and an improved immune system
- Fresher breath, less bloating, reduced body odour and fewer headaches
- A reduction in cholesterol and high amounts of saturated fats (found in animal-based foods)
- A wider range of nutrients, vitamins and minerals
- Fewer food cravings and less snacking
- No exposure to growth hormones or bacteria that are often found in animal products
- Lower rates of heart disease, diabetes, cancer and obesity among those who don't consume animal products

*Source: Springmann et al., "Health and nutritional aspects of sustainable diet strategies and their association with environmental impacts: a global modelling analysis with country-level detail", *The Lancet Planetary Health* (October 2018).

MAKING THE CHANGE

Changing to a plant-based or plant-focused diet has never been easier. There are plenty of vegan products available in supermarkets, restaurants offer vegan alternatives, and airlines and train companies have vegan options. It is a great time to make the change. Plus, as more vegan products hit the shelves, prices are dropping. Remember that a vegan diet is based around fruit, vegetables and grains – these are generally inexpensive ingredients, especially when seasonally and locally grown.

For most people, a gradual change works best. Start by replacing one daily meal, such as breakfast, with a plant-based alternative. Swap recipes with friends. Experiment with alternatives to the dishes you already enjoy. Take the time to learn to cook your own vegan feasts, rather than relying on ready-made vegan meals. Finally, do not expect to transform your diet in one fell swoop – enjoy the journey.

THE RESPONSIBLE CONSUMER

"Never doubt that a small group of thoughtful, committed citizens can change the world. Indeed, it is the only thing that ever has."

Margaret Mead

AWARENESS IS KEY

Being a responsible consumer is not only about being "green". It is also about considering the social impact of our food choices. This means being aware of where in the world our food comes from and *who* is producing it.

In this chapter, we will discuss the produce that is planet-friendly and healthy for us, alongside products that are best to avoid due to their detrimental impact on both the environment and the communities who struggle to produce them.

Taking this knowledge with you during every food shop will allow you to gradually and naturally change the contents of your shopping baskets to more ethical, planet-friendly products. If we all do this, imagine the long-term, positive impact we could make.

SAVVY SHOPPER CHEAT SHEET

Top five cooking oils

1. Rapeseed
2. Sustainable vegetable or sustainable palm oil
3. Sunflower
4. Coconut
5. Olive, organically produced

Top five water-efficient salad ingredients

1. Tomato
2. Lettuce
3. Sugar snap pea
4. Pumpkin/squash
5. Carrot

Top five staple foods

1. Potato and sweet potato – sustainable in most climates
2. Wholegrain wheat – sustainable in cooler regions with average rainfall
3. Oats – sustainable in most cooler regions
4. Corn – sustainable in regions with high rainfall
5. Rice – sustainable in tropical regions with high rainfall

Top five planet-friendly home-made snack ideas

1. Carrots, cut into sticks
2. Cinnamon apple chips (see recipe on page 68)
3. Boiled eggs, organic and locally sourced
4. Home-made jam on wholemeal or buckwheat toast
5. Kale chips (see recipe on page 116)

Spice up your life

Once upon a time, herbs and spices were mostly grown in South and South-East Asia. Today, they are commonplace all over the world. Choose organic and fair-trade herbs and spices to ensure they have been produced in the most ethical and planet-friendly way possible. Even better, opt for brands using ingredients grown as locally as possible.

Remember!

The more natural the food product, the less energy required in the production process. For example, fresh tomatoes are more planet-friendly than tinned tomatoes, but the latter keep for longer (and the tin can be recycled).

Avoid the bandwagon

An unsustainable demand for a specific product can lead to the abandonment of sustainable farming practices, local price hikes and the exploitation of farm workers in regional communities. This not only has a devastating effect on the local environment, but also impacts the well-being of under-pressure farm workers. We can help by being more mindful of our consumption of imported "on-trend" foods, such as chia seeds, açai, avocado and quinoa (see pages 42-43).

TO BUY OR NOT TO BUY: TROPICAL FRUITS?

Exotic fruits – pineapples, mangos and lychees, for example – are transported thousands of miles to reach us, meaning their carbon footprint is high. Long-distance distribution also often means more energy is used in refrigeration to keep fruits fresh, and excess packaging is needed to keep them protected. Enjoy tropical fruits, but with more awareness. For example, consume them less regularly or save them for a treat when holidaying in warmer climates.

Know your bananas

Nutritious and high-yielding, bananas are loved across the world. They provide us with much-needed potassium – and are a great source of energy. However, the intensive farming of bananas has led to soil degradation, water pollution, deforestation and the mistreatment of workers. Not to mention the greenhouse gas emissions that come from the long-distance transportation! The Rainforest Alliance assists banana farmers in meeting their environmental, social and economic sustainability standards, ensuring planet-friendly practices are adhered to – look for the stamp featuring a green frog.

THE BENEFITS OF BUYING LOCAL

Locally grown or sourced food is a great way to be planet-friendly and healthy. By buying local, we are:

- Enjoying fresh and flavoursome seasonal produce
- Eating produce picked at the peak of its ripeness
- Supporting the local economy and small businesses
- Eating more nutritious food due to minimal time between harvest and consumption
- Preserving the natural genetic diversity of plant crops
- Ensuring a safer supply due to less processing and handling
- Inspiring producer accountability for food quality
- Reducing shipping and transport distances, resulting in a lighter carbon footprint
- Rewarding local growers and distributors with fair prices, cutting out high costs of storage, packaging and transportation of long-distance food

FEEDING YOUR SOUL

A common misunderstanding associated with a planet-friendly diet is that it means giving up delicious, nutritious and varied meals. Thankfully, this is not the case. Remember, a plant-based diet should still provide us with balanced, nourishing, affordable options. These top crops will keep you full, replenish nutrients *and* contribute to a healthy environment.

Peas, beans and lentils (pulses)

For you: Full of fibre, protein and vitamins; help to reduce the risk of type 2 diabetes

For the planet: Nitrogen-fixing crops capture nitrogen from the air and convert it into nutrients for the soil. They also require less water and synthetic fertilizer

Cherries

For you: Low-calorie snack; contain fibre, calcium and potassium

For the planet: Cherry trees provide a habitat for birds and local wildlife

Sweet potato

For you: Good source of iron, calcium, fibre and antioxidants; great for gut health; filling

For the planet: High-yielding crops; produces low carbon emissions and requires less water

Tomatoes

For you: Great source of vitamin C, potassium and vitamin K

For the planet: Produce low carbon dioxide emissions during production; require less water; easy to grow at home

Buckwheat

For you: Gluten-free alternative to wheat; high in protein and fibre, improving heart health and reducing blood-sugar levels

For the planet: Cultivated across the world; attracts insects and pollinators to blossoms

Mushrooms

For you: Protein-rich, full of fibre and iron, and low in sodium; contain vitamin B, which is vital for a healthy nervous system; white button mushrooms are one of the few sources of vitamin D not associated with animal products

For the planet: Edible mushrooms are sustainably grown in most climates

---

The World Health Organization advises that a daily diet rich in fruit and vegetables can help reduce the risk of heart disease, strokes and some cancers, and guard against unhealthy weight gain.

RED-LIGHT FOODS

The production of some food has a devastating impact on environments or communities – sometimes both. These products are not recommended...

Avocados

A large proportion of avocados come from Central America, where a booming demand has caused prices to skyrocket and the staple food to become unaffordable locally. Intensive farming has led to forests being thinned.

Prawns (and shrimp)

It is estimated that one third of the world's discarded by-catch comes from prawn fisheries. Trawlers catch 10–20 kilograms of by-catch, including turtles, while collecting just one kilogram of prawns.[*] In the tropics, mangrove forests are destroyed to create prawn farms. Tip: If you must buy prawns, buy sustainable, cold-water prawns.

Sugar from sugar cane

Forests across the world have been cleared for sugar cane plantations. The crop is particularly susceptible to disease, meaning farmers rely heavily on fertilizers and pesticides. Tip: Sugar from locally farmed sugar beets is a sustainable alternative – check the label. Local apiary (beehive collective) honey is another great option.

[*]Source: Food and Agriculture Organization of the United Nations.

Quinoa

Hailing from the cool, arid mountains of Peru and Bolivia, quinoa has become a lucrative earner for farmers. Unfortunately, demand has lured in intensive farming methods, including heavy use of fertilizers and pesticides, and exploitative labour.

Nuts

Nut crops require large-scale irrigation – using up huge volumes of water for a small amount of produce – and that is before the harvest is transported long distances. Almonds, hazelnuts, walnuts and Brazil nuts are the least sustainable. Tip: Peanuts are nitrogen-fixing legumes and may be grown sustainably on small farms.

Chocolate

The key ingredients in chocolate – cocoa beans, milk powder and butter (dairy), sugar (sugar cane), eggs (chickens) and palm oil – are often produced using unsustainable methods. Cocoa is predominantly grown in West Africa, and tropical forests are still being cleared to meet plantation demands. Large-scale irrigation is necessary. As demand increases, so does child labour. Tip: Organic, fair-trade chocolate provides a kinder alternative, keeping workers and their welfare in mind.

The Ivory Coast's loss of more than 80 per cent of its forests over the last 50 years is due mainly to cocoa production.*

*Source: World Wildlife Fund.

THE PROBLEM WITH PALM OIL

Palm oil is a vegetable oil made from the kernel of the oil palm tree. It is an ingredient found in approximately 50 per cent of packaged food (and many other non-edible products as well). Plantations in Indonesia and Malaysia produce 85 per cent of the world's palm oil – and are a prominent cause of deforestation.* As a result, wildlife species, such as the Sumatran and Bornean orangutan, are critically endangered. Rural communities have also been forced to leave their homes.

The Roundtable on Sustainable Palm Oil (RSPO) has certified certain products. Look for their green label. To claim sustainability, manufacturers must pledge:

- Not to clear primary rainforests or critical ecosystems
- To limit planting on peatland
- To protect, conserve and enhance ecosystems, wildlife and natural resources
- To respect human rights, treat workers fairly and deliver local benefits
- To behave ethically and enforce transparent supply chains
- To operate legally, with efficiency and to optimize productivity

Look out! Sometimes palm oil has other names too...

Vegetable oil, vegetable fat, palm kernel, palm kernel oil, palm fruit oil, palmate, palmitate, palmolein, glyceryl, stearate, stearic acid, sodium palm kernelate, octyl palmitate, palmityl alcohol, to name a few... You may also see "crude" palm oil (which is made from the tree's fruit, rather than the kernel).

*Source: World Wildlife Fund.

DO!

- Eat at farm-to-table restaurants, where chefs make a conscious effort to use sustainably grown, locally sourced ingredients
- Support local businesses that do not use single-use plastic packaging
- Opt for locally sourced wines to reduce pollution from transportation or choose beer on tap over bottled beer to reduce waste
- Visit eateries that serve seasonal and local produce
- Choose eateries with a menu detailing where their food is sourced from
- Take uneaten food home to eat later
- Always carry a reusable cup to take away hot drinks

DON'T!

- Go to chain restaurants or fast food outlets that mass-produce food with little consideration given to its source or ingredients
- Buy from places that overuse single-use plastic packaging
- Order bottled water; instead request tap or filtered water

Studies suggest that over 80 per cent of consumers consider sustainability when deciding where to eat out. One third sampled said they would be willing to pay more for a planet-friendly dining experience.*

*Source: The Sustainable Restaurant Association.

MAKING SENSE OF FOOD LABELS

Most pre-packaged food has an ingredient label on the back or side of the packaging. The label also typically shows where your food has been sourced from, if it is approved by a recognized sustainability body and whether it is organic. These are a few common terms you may find on labels:

"ORGANIC" – at least 95 per cent of ingredients are grown organically and the farm or farms have been certified by a registered organic control body

"FREE RANGE" – animals and birds should have some access to outdoor space

"WILD" – seafood from a wild marine environment

"VEGAN" – contains no meat, poultry, fish, dairy or any ingredient of animal origin

"NON-GMO" – food that hasn't been genetically modified

"SUGAR FREE" – less than 0.5 grams of sugar per serving

"FAIR TRADE" – certified in accordance with fair-trade standards

TRUST THE SUSTAINABILITY STAMP

Look out for the following stamps that certify that a product meets international sustainability standards.

- A blue Marine Stewardship Council (MSC) label certifies seafood from sustainable wild fisheries
- The turquoise Aquaculture Stewardship Council (ASC) label certifies seafood from sustainable and socially responsible producers
- Distinguished by its blue, green and black stamp, the Fairtrade Foundation seal confirms ingredients are sourced and traded under ethical conditions
- Palm oil from a credited plantation is marked with a Roundtable on Sustainable Palm Oil (RSPO) stamp
- The Rainforest Alliance's green frog seal certifies a product sourced from farms with high environmental, social and economic sustainability standards

Local stamps that also provide assurance

- "RSPCA Assured" guarantees that UK-farmed meat is from higher-welfare farms
- Certified organic produce from Europe is marked with the Soil Association's black and white "Organic Standard" stamp
- The US Department of Agriculture's green and white "USDA Organic" label ensures meat and crops are 100 per cent organically produced

THE PLANET-FRIENDLY SHOPPING LIST

Now that we have an idea of the food products and practices that promote the well-being of the planet, and those that can diminish it, why not implement some changes for your next food shop?

How many points can you tick off this list?

☐ Is your basket composed of mostly fruit and veg?

☐ Are the fruits and veggies in season?

☐ Are they grown locally?

☐ Is there a sustainability stamp or label on the meat or fish?

☐ Have you chosen loose fruits and vegetables?

☐ Is there little or no packaging?

☐ Is the seafood listed as locally sustainable on the Marine Conservation Society Good Fish Guide app?

☐ Will you consume this product by the use-by date?

☐ Could you use plant milk instead of dairy?

☐ Can you pick veggies that aren't aesthetically perfect? They taste the same!

☐ Can you buy any items from a farmer's market? If so, do

☐ Does this product contain palm oil? Look for the "RSPO" stamp

☐ Can you choose fresh produce and avoid processed and heavily packaged versions?

GO FORAGING

There is something empowering about going in search of and gathering your own food. Foraging allows you to pick natural, local plants – including herbs and fruits – and turn them into delicious meals and snacks. It also helps the planet in the process, as you're reducing your reliance on packaged food transported over long distances. If you are tempted to try foraging, here are the three main rules to follow:

1. Know the legalities

Is it legal to forage in the area? Check local authority websites for up-to-date policies – remember that some plant species are protected by law.

2. Know what to pick

Be sure of what you are picking. Carry reference books for identification, though completing a local foraging course is better. Never pick or eat plants without being certain of their identity. Mushrooms and berries are particularly hard to identify and can be poisonous. If in doubt, do not eat them.

3. Know how to forage sustainably

Wild plants are food to local wildlife, so be mindful and take only what you need. Avoid damaging roots and ensure there are enough plants for species regeneration. Never forage in conservation areas.

Sign up for a local foraging course and know your species before you start!

THE ENVIRONMENTALLY FRIENDLY KITCHEN

"Food – from what we grow, produce and catch to what we put on our plates – is the hidden cause of biodiversity loss... Reconnecting people with food is essential if we want to... build a future of healthy people and a healthy planet."

Sarah Halevy, Sustainable Diets Manager, World Wildlife Fund UK

EFFORTLESS KITCHEN HACKS

Establishing a planet-friendly kitchen is easy when you know how. Here is how to make the most of your produce, save energy and reduce food waste...

Top ten food storage tips

1. Place a paper towel in salad bags to soak up moisture and prevent spoiling
2. Store potatoes in a cool, dry place to prevent fast sprouting
3. Keep citrus fruits and apples separate from other fruits as they can cause them to ripen quickly
4. Don't throw out overripe bananas – they are delicious in banana bread and smoothies or frozen and blended into a dairy-free ice cream
5. Keep grains in airtight containers to maintain freshness
6. Garlic will last longer if stored in an open basket or paper bag
7. Store dairy at the back of the fridge, where the temperature is cooler
8. Keep meat on the bottom shelf of the fridge to avoid juice dripping down and spoiling other foods
9. Keep cheese fresh by wrapping in a porous material to allow air in and moisture out
10. Fill your freezer. It will operate with more energy efficiency when full

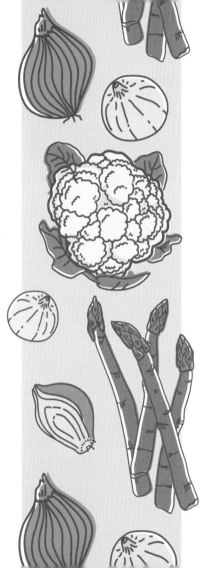

THE FOOD WASTE DILEMMA

Between 1 and 2 billion tonnes of food is wasted annually. That is around 30 to 40 per cent of our food.* And it isn't just the product that ends up in the trash – the water and energy used in the production process also goes to waste. In fact, the equivalent of 552 million Olympic-sized swimming pools' worth of water is lost annually through wasted food.**

Food is wasted when:

- Edible fruit and vegetables don't meet strict physical characteristics set by supermarkets – usually size, shape and colour
- It is damaged during transport and handling
- Incorrect use-by dates are assigned
- It remains uneaten after overbuying

The amount of food wasted each year is enough to feed poverty-stricken communities around the world, several times over.

*Source: Food and Agriculture Organization of the United Nations.
**Source: *National Geographic*, "365 trillion gallons of water thrown away with our food every year" (September 2010).

TIPS TO REDUCE FOOD WASTE

The art of planning ahead

A meal plan is a simple and effective way to prevent food waste:

1. Decide how many meals you will enjoy at home over the week, factoring in takeaways and leftover nights

2. Select planet-friendly recipes based on the time you have to cook. Choose meals that:
   - You already love
   - Share common ingredients
   - Are big enough to allow leftovers you can store and eat another day

3. Note the ingredients. Then cross off what is already in your cupboards

4. Add basic planet-friendly ingredients such as rapeseed oil, unsweetened plant milk, potatoes, tomatoes, and dried herbs and spices. Finalize the list

5. Shop only for what you need to avoid overbuying

What a pickle!

Developed as a method of preservation, pickling prevents food from spoiling quickly and allows storage for months if refrigerated. Find out how to pickle veggies on page 110.

Shop at local markets for fresh fruit and veg that don't meet the aesthetic requirements of supermarkets.

TOP TEN ENERGY-EFFICIENT COOKING TIPS

1. Consume fruits and veggies in their natural state where possible – for example, rather than roasting carrots, grate them into a salad
2. Soak pasta and pulses before cooking to soften, reducing cooking time
3. Save on energy by steaming veggies in a colander over a hob already in use
4. Defrost produce naturally
5. Use microwaves for heating – they use less energy than stoves
6. Electric hobs are more energy-efficient than gas hobs. Boil water in a kettle rather than in a pan – it uses less energy
7. Turn off appliances – toasters, kettles and microwaves – at the plug when not in use
8. Use the appropriately sized pan for the volume of water you need
9. Ovens use more energy than hobs – batch-bake to avoid daily use
10. Cook as much as possible in one swoop – and freeze or refrigerate the leftovers

THE ROAD TO SELF-SUSTAINABILITY

Growing your own food is a huge step toward achieving a more planet-friendly diet. You might start off with a small pot of herbs growing by your kitchen window, or a vegetable patch in your garden. By growing your own produce, you are not only producing food in a local and organic environment, but you are also saving on energy, transport emissions and packaging – as well as helping to reduce the demand on the food production industry.

Most vegetables can be grown from seeds or seedlings. Planting successive crops means you can harvest vegetables throughout the year, allowing for a seasonal, varied and healthy food supply. It is also immensely satisfying and inspiring to grow your own food from scratch.

START YOUR OWN VEGGIE PATCH

- Pick somewhere with ample sunlight
- Start with a small space to master the basics
- A raised garden bed is ideal. Planter boxes are equally suitable
- Allow 30 to 45 centimetres of soil depth for plants to take root, and plenty of drainage for excess water
- Buy a peat-free, organic compost that doesn't contain chemicals or artificial fertilizers – better still, make your own (see below)
- Add a layer of mulch – grass cuttings, shredded newspaper or leaves – on top of the soil to retain moisture, control temperature, suppress weeds and provide nutrients
- Water raised beds twice a week and planters every other day, unless you live somewhere with an extreme climate

Make your own compost using leaf litter, kitchen scraps, eggshells, shredded newspaper, straw and grass cuttings. Store in a box, adding to the pile when possible. Keep damp and away from light, stirring weekly to allow oxygen in. Once dark in appearance, the compost is ready.

Herb garden

Grow coriander, parsley, basil and mint in full sunlight and moisture-retentive soil, with plenty of compost. Rosemary and thyme are perennial herbs and will regenerate each year.

WHAT TO PLANT? FOOD FOR ALL THE SEASONS

Plant in winter for a spring harvest

Lettuce – one seed packet can provide five months of harvest

Broccoli – grows well in cooler climates; allow 30 centimetres between seeds

Spring onion – quick-growing

Carrots – plant in deep, loose soil; harvest when the flowers appear

Plant in summer for an autumn harvest

Potatoes – high-yielding; fast-growing

Peas – easy to grow; require minimum space

Cabbage – single crop; thrives in light frost; mulch well

Rocket (arugula) – fast-growing; yields until spring

Plant in spring for a summer harvest

Tomatoes – require minimum space; easy to grow

Courgettes – grow indoors (if cool) or outdoors (in warmer climates)

Sweet peppers – pinch off taller shoots for a bushy plant

Squash and pumpkin – grow from seed in a sunny, wind-sheltered spot; cover until germination

Plant in autumn for a winter harvest

Spinach – plant in batches for a continuous yield

Parsnips – allow deep pockets for roots; keep soil weed-free

Kale – continuous harvest over winter

Asparagus – grow in sunshine; keep soil weed-free

CREATING A NATURE-FRIENDLY GARDEN

- Introduce bird boxes, feeders and water baths to attract birdlife
- Plant flowers, such as lavender and bee balms, to draw in bees and butterflies
- Climbers like ivy, clematis and roses provide cover for birds and insects
- Allow wild flowers and grass to grow
- Home-made compost is great for worms and other decomposers
- Rocks, twigs and rotting wood provide insect shelters
- Create a pond using a buried bucket or trough; add stones and twigs to help wildlife to enter and exit

Keeping the pests away

- Watch out! Inspect plant crops regularly
- Pests include slugs, caterpillars and aphids
- Hand-pick any pests rather than use synthetic pesticides
- Ladybirds, lacewing flies, frogs, hedgehogs and birds are great at natural pest control
- Grow "sacrificial" or "trap" plants, such as red clover or marigolds, to draw slugs away from food crops
- Mint, basil and rosemary are also natural pest deterrents and can be grown in among veggies

DIY GARDEN AIDS

Produce your own organic pest deterrent

Ingredients: 1 garlic head, 500 ml water,
1 tbsp sustainable vegetable oil,
1 tbsp eco-friendly dishwashing liquid

1. Purée peeled garlic cloves with water and oil, and leave overnight
2. Strain and add dishwashing liquid
3. Mix
4. Pour into a spray bottle and use on plants

Make your own fertilizer

Ingredients: vegetable scraps,
Epsom salts, water

1. Store scraps in the freezer until you have a large bag
2. Thaw scraps and blend to smooth consistency
3. Pour purée into bucket
4. Add ½ tsp of Epsom salts for every blenderful of purée
5. Stir and allow to sit overnight
6. Combine 1 litre of purée with 3.5 litres of warm water
7. Mix
8. Apply to the base of plants

Part 2

PLANET-FRIENDLY RECIPES

 vegetarian vegan  gluten-free

HUEVOS RANCHEROS

Start your day with a burst of energy with this oh-so-good traditional Mexican breakfast. Throw in extra garlic and chilli for a spicy kick and serve with a healthy portion of refried beans (page 106).

Serves: 2 · Preparation: 10 mins · Cooking: 15 mins

Method

Heat the oil in a frying pan over a medium heat. Add the red pepper and cook until soft. Introduce the spring onion, garlic, chilli and oregano, and cook for 2 more minutes, then add the tomatoes and allow to simmer for 5 minutes. Preheat the grill to the highest setting. Pour the pepper mix into a shallow ovenproof dish. Make two spaces in the middle using a spoon, and crack in the eggs. Season with salt and pepper to taste. Cook under the grill for 5 to 6 minutes, or until the eggs are cooked to preference. Meanwhile, toast the pitta bread. Serve the piping hot huevos rancheros on the pitta.

Ingredients

1 tbsp rapeseed oil

1 deseeded, chopped
red pepper

115 g chopped greens
of spring onion

1 peeled, sliced
garlic clove

½ sliced green chilli

1 tsp oregano

2 chopped tomatoes

2 organic eggs

2 wholemeal
pitta breads

Salt and pepper

EGGY BREAD AND VEGGIE MUFFINS

This impressive breakfast recipe is fantastic for using up any old bread you have lying around. Swap in your favourite (locally sourced) root veggies and greens to liven up your early-morning repertoire.

Serves: 2 · Preparation: 10 mins · Cooking: 45 mins

Method

Preheat the oven to 180°C/350°F/gas mark 4 and line a muffin tray with folded cups of greaseproof paper. Ensure the veggies are cut into small pieces, then fry them in oil until brown. When cooked, stir in the spinach until it wilts. Remove from the heat.

Meanwhile, chop the stale bread into cubes and place in a bowl. Whisk the eggs and milk together and pour half the mixture on to the bread. Stir and set aside for 10 minutes. Once thick, stir the vegetables through the mixture and season to taste. Pour into the muffin cases and top with the remaining eggy bread mix. Bake until golden and crispy (approximately 30 minutes). Serve warm.

Ingredients

1 sliced courgette

½ deseeded, diced yellow pepper

100 g chopped butternut squash or pumpkin

250 g shredded spinach

1 tbsp rapeseed oil

8 slices stale wholemeal bread

4 organic eggs

150 ml unsweetened plant milk

Salt and pepper

HOME-MADE HONEY GRANOLA

Making your own granola means you can choose wholly sustainable ingredients for your breakfast. This recipe fills a medium jar and will be plenty for four to five breakfasts.

Makes: 1 medium jar · Preparation: 10 mins · Cooking: 50 mins

Method

Preheat the fan oven to 160°C/325°F/gas mark 3 and line a large tray with baking paper. Heat the oil and honey in the microwave – it is less energy-consuming than on the stove – until melted (approximately 30 seconds), then add the cinnamon and a squeeze of orange juice. With the exception of the dried fruit, add the remaining ingredients into a bowl, followed by the oil and honey, and mix well. Then spread the complete mix evenly over the tray. Bake for 45 minutes, or until golden, stirring every 4 to 7 minutes. Once browned, remove from the oven and leave to cool. Finally, add the dried fruit and store in an airtight jar. Enjoy with plant-based yoghurt and cinnamon apple chips (page 68).

Ingredients

6 tbsp coconut oil

150 ml organic honey

4 tsp ground cinnamon

1 peeled orange

400 g organic oats

70 g sunflower seeds

50 g pumpkin seeds

100 g locally sourced, seasonal dried fruit

CINNAMON APPLE CHIPS

These cinnamon apple chips are ideal as a fruit topping on granola or porridge, for an on-the-move breakfast box or for those who nibble throughout the day. This is a simple fruity recipe that can be batch-made and stored for up to a week.

Makes: 1 medium jar · Preparation: 5 mins · Cooking: 2–2½ hours

Method

While preheating the oven to 140°C/275°F/gas mark 1, cut the apples in half and remove the cores. Then slice into pieces, approximately 5 to 7 mm thick, and place into a bowl. Stir in the cinnamon, coating the apples thoroughly. Spread the apples over a wire oven rack, making sure nothing overlaps. Insert the rack in the middle of the oven and bake for approximately 1 hour. Once the hour is up, rotate the rack. After another hour, test the chips by removing one, allowing to cool and checking whether it is crispy enough. If still soft, continue to bake, checking the chips every 15 minutes until ready. Once out of the oven, allow the chips to cool down completely. Then transfer them to an airtight container to be stored.

Ingredients

2 kg red apples

3 tsp ground cinnamon

SPINACH-AND-MUSHROOM-STUFFED BUCKWHEAT PARCELS

This delicious gluten-free dish is a great treat to expand your breakfast options. Use in-season, locally grown mushrooms and serve with MSC-certified smoked salmon as an optional extra. Soya cream is most readily available; however, it can be replaced with any plant-based cream of choice.

Serves: 2 · Preparation: 20 mins · Cooking: 30 mins

Ingredients

40 g buckwheat flour

3 organic eggs

125 ml plant milk

1 tsp Dijon mustard

2 tbsp soya cream

1 tbsp organic olive oil

150 g sliced mushrooms

2 sprigs thyme

80 g steamed spinach

Salt and pepper

Method

Preheat oven to 180°C/350°F/gas mark 4 and line a baking tray with foil. Whisk the flour, an egg and milk in a bowl. Mix the mustard and cream in another bowl. Fry the oil, mushrooms and thyme together until golden. Pour half the buckwheat mix into a non-stick frying pan and swirl into a crêpe-thin layer. Cook until brown, flip and brown the other side. Take off heat and spoon half the mustard and cream mix over the pancake. Crack an egg into the centre and surround with half the spinach. Fold the sides of the pancake over the spinach, framing the egg, then top with half the mushrooms. Heat for another minute, then transfer to a baking tray. Repeat the process. Bake until the egg whites set. Season to taste and serve.

HADDOCK FISHCAKES

Unbelievably succulent, full of nutritional goodness and simple to make, these fishcakes are a fabulously filling, healthy lunch option – and can be made using any locally caught sustainable fish. Serve with boiled beetroot as an optional extra.

Serves: 2 · Preparation: 10 mins · Cooking: 30 mins

Method

Chop the potato into small chunks and boil until tender (approximately 15 minutes). Once ready, drain and mash to a smooth consistency, then place in a large bowl. Poach the fish in a shallow saucepan of water for 5 minutes; drain and allow to cool. Break the fish into small pieces using your hands and add to the mashed potato. Mix using a wooden spoon, stirring in the parsley, egg and flour. Lightly season. Shape into four patties and fry in the oil, until golden on each side. Serve with rocket and a wedge of lemon, and with the boiled beetroot, if desired.

Ingredients

250 g peeled potato

2 fillets MSC-certified skinless and boneless haddock

1 peeled and cubed boiled beetroot (optional)

1 tbsp chopped parsley

1 organic egg

1 tbsp wholewheat flour

Salt and pepper

2 tbsp rapeseed oil

60 g rocket

½ lemon

CREAMY MUSHROOMS
SERVED ON CRUSTY BREAD

A great source of vitamins, this healthy mushroom offering is perfect for a lunchtime filler or as a light dinner option. Use button mushrooms for a more delicate taste.

Serves: 2 · Preparation: 10 mins · Cooking: 20 mins

Method

Drizzle a pan with olive oil and cook the onion and garlic on a medium heat, until brown and soft. Stir in the mushrooms and leave the mix to gently fry for 5 to 10 minutes, until well cooked. Spoon in the yoghurt and lemon juice and stir well, before adding the pinch of salt to taste. Leave on the heat for 3 to 4 minutes, while toasting the bread. Once ready, present the mushrooms on the sourdough toast with a sprinkle of thyme and pepper.

Ingredients

2 tbsp organic olive oil

½ onion, peeled and chopped

6 garlic cloves, peeled and chopped

500 g locally sourced chestnut mushrooms, chopped

2 tbsp coconut yoghurt

1 lemon, juiced

Pinch of sea salt

4 slices sourdough

4 sprigs fresh thyme

Black pepper

COURGETTE FRITTERS

Simple, quick to prepare and versatile, this veggie offering is great for a late-morning pick-me-up. Boost the meal into lunch with an optional serving of poached organic eggs or even some sustainably sourced smoked salmon – delicious!

Serves: 2 · Preparation: 15 mins · Cooking: 10 mins

Method

Grate the courgette and allow the moisture to drain by placing on a clean dish towel for 10 minutes. Combine the courgette, flour, garlic and egg in a bowl with the salt and a pinch of pepper. Heat the olive oil in a frying pan over a high heat. Spoon the fritter mix into the pan and flatten with a spatula. Cook until golden brown underneath and then flip to cook the other side (approximately 4 minutes in total). Serve on its own, on toast or with the optional poached egg or smoked salmon.

Ingredients

1 courgette

30 g wholewheat flour

1 peeled garlic clove

1 free-range organic egg

1 tsp salt

Pinch of pepper

2 tbsp organic olive oil

SWEET POTATO, CARROT AND GINGER SOUP

This warming soup, packed with vitamin-heavy veggies, is an ideal lunch for a cold day. It's simple, delicious, healthy and planet-friendly – batch-cook and then refrigerate the leftovers to enjoy throughout the week.

Serves: 2 · Preparation: 10 mins · Cooking: 35 mins

Method

Add the oil, sweet potato, carrots and onion to a large pan and cook for 8 minutes. Add the ginger and cook for a further 3 minutes before adding the stock. Simmer the mix for a further 20 minutes, until the sweet potato and carrots are tender to touch with a fork. Turn off the heat and transfer everything into a food processor, pulping until smooth. If the consistency is too thick, feel free to add a little water. Season with salt and pepper to taste, garnish with parsley and serve in a bowl.

Ingredients

1 tbsp organic olive oil

250 g chunks of peeled sweet potato

150 g chopped carrots

½ diced onion

1 finely grated ginger root (approx. 4 cm)

750 ml vegetable stock

Salt and pepper

A few parsley leaves

FRESH HERB AND MUSHROOM PASTA

This easy-to-create lunch is ideal for those weeks when time is of the essence – even better, leftovers can be stored for a second day. Throw in extra seasonal greens to boost the veggie goodness and a few extra cloves for an enhanced garlic flavour.

Serves: 2 · Preparation: 15 mins · Cooking: 30 mins

Method

Boil the pasta until well cooked, remembering that gluten-free pasta may take a little longer than regular pasta. Drain using a colander and leave to cool. Drizzle the oil in a pan and fry the onion and garlic until soft and golden. Add the mushrooms and cook for approximately 10 minutes. Stir in the spinach and allow to wilt. Once cooked, mix in the herbs, yoghurt and pasta, squeezing in the juice of the lemon. Add salt and pepper to taste.

Ingredients

150 g wholegrain or gluten-free pasta (any)

1 tbsp organic olive oil

1 peeled and chopped onion

3 peeled and chopped garlic cloves

200 g sliced button mushrooms

Large handful spinach

Small handful sage leaves

Large handful fresh basil

2 tbsp unsweetened soya yoghurt

½ lemon

Salt and pepper

HERB-CRUSTED COD AND POTATO SALAD

A combination of fresh, sustainably sourced ingredients ensures this fish dish is bursting with delicate flavours and remains planet-friendly. Swap the cod for any sustainably caught white fish that is local to you.

Serves: 2 · Preparation: 15 mins · Cooking: 35 mins

Method

If you're serving with potatoes, bring a small saucepan of water to the boil and add the potatoes and basil. Once wilted, remove the basil and leave to cool. When softened, drain the potatoes and keep to one side. Combine the chives, parsley, carrot, breadcrumbs, pepper and a pinch of salt in a bowl and mix. Brush the fillets with oil and place on a baking tray. Scatter the herbs over the fish to create a crust. Bake in the oven at 220°C/425°F/gas mark 7 for 15 minutes. Meanwhile, blend the basil, olive oil, vinegar, garlic and salt in a food processor to make a dressing. Serve the fish drizzled in dressing and garnished with the lemon and parsley leaves. Add a side of boiled potatoes, if desired.

Ingredients

600 g halved baby potatoes (optional)

Bunch of basil

2 tbsp chopped chives

3 tsp dried parsley

1 peeled and grated carrot

40 g breadcrumbs

½ tsp black pepper

2 pinches of salt

2 MSC-certified cod fillets

100 ml organic olive oil

2 tbsp white wine vinegar

1 peeled garlic clove

2 lemon wedges

Handful of chopped parsley

VEGGIE FAJITAS

A speedy option for a last-minute dinner party, this delicious fajita mix can be spiced up to suit your taste buds. Pickle your own jalapeños and take this planet-friendly recipe to an even higher level of sustainability.

Serves: 2 · Preparation: 15 mins · Cooking: 15 mins

Method

Mix the salsa ingredients together in a bowl and leave to infuse. Fry the onion and pepper in a pan, using the oil, allowing them to become brown and soft. Lower the heat, then add the spices and a large pinch of salt. Stir in the beans with 1 tbsp of water and leave to cook until the beans are steaming hot and the water has reduced. Once ready, serve the fajita mix on to warm tortillas. Top with the salsa.

Ingredients

For the fajitas:

1 sliced onion

1 sliced red pepper

1 tbsp rapeseed oil

½ tsp ground cumin

¼ tsp mild/hot chilli powder

¼ tsp hot paprika

Large pinch of salt

1 tin kidney or black beans, rinsed

4 flour tortillas

For the salsa:

2 diced tomatoes

¼ chopped red onion

2 tsp chopped pickled jalapeños

½ lime

Handful of chopped coriander

RAINBOW ROAST VEGETABLE SALAD

A wholesome side salad or a hearty vegan main, this colourful platter is a versatile option for any meal. Use locally grown pumpkin or butternut squash – and don't forget to save unused seeds for a future veggie garden (or toast them and eat them as a healthy snack).

Serves: 2 · Preparation: 20 mins · Cooking: 40 mins

Method

Preheat oven to 190°C/375°F/gas mark 5. Scoop out the seeds and save, before cutting the pumpkin or squash into small chunks. Lay the pieces in a roasting tray. Scrub the beetroots, parsnips and carrots, before also chopping into chunks and adding to the tray. Slice the onion into thin wedges and tuck in between the root veggies. Cut the fennel bulb into six wedges and place on the tray. Drizzle the vegetables with olive oil. Crush the coriander seeds using a pestle and mortar then sprinkle over the veggies, season with salt and pepper to taste, and toss. Roast the vegetables for 40 minutes, stirring halfway through, until they soften and become golden in colour. Remove and allow to cool. Add a few mint leaves and parsley and serve.

Ingredients

¼ peeled pumpkin or butternut squash

3 beetroots

2 small parsnips (optional)

2 carrots (optional)

½ peeled red onion

1 fennel bulb

Drizzle of organic olive oil

2 tsp coriander seeds

Salt and pepper

Fresh mint leaves

Flat-leaf parsley

CHILLI PRAWNS WITH CAULIFLOWER RICE

Enjoy this tasty, light and sustainable recipe. Swap the veggies for seasonal preferences where needed and be sure to use Marine Conservation Society-certified, locally caught, cold-water prawns for this mouth-wateringly yummy dish.

Serves: 2 · Preparation: 10 mins · Cooking: 20 mins

Method

Heat half the oil in a pan at medium heat, cooking half the garlic, ginger and chilli for 1 minute. Stir in the prawns and cook through until white in colour, then transfer the mix into a bowl. Blend the cauliflower in a food processor until fluffy. Cook the spring onions, pepper and courgette in the remaining oil for 3 minutes, then add the cauliflower, remaining garlic, ginger and chilli. Cook for 5 minutes. Mix the dressing ingredients in a bowl. Serve the prawns on top of a bed of the cauliflower mix, drizzled with dressing and garnished with coriander or mint.

Ingredients

1 tbsp sustainable vegetable oil

1 finely chopped garlic clove

Grated ginger root (approx. 2 cm)

1 deseeded red chilli

150 g peeled, raw, MSC-certified cold-water prawns

300 g cauliflower florets

5 chopped spring onions

1 sliced red pepper

1 courgette, peeled into ribbons

1 tbsp chopped fresh coriander or mint

For the dressing:

1 tsp organic honey

1 lime, juiced

1 tsp organic olive oil

1 tbsp soy sauce

PEPPER, MUSHROOM AND HERB PIZZA

A tasty vegan pizza that is full of nutritional goodness. Particularly water-efficient veggies have been chosen for the pizza topping, but other sustainable alternatives, such as root vegetables and leafy greens, also work perfectly.

Serves: 2 · Preparation: 15 mins · Rising time for the dough: 1 hour · Cooking: 25 mins

Method

Mix the flour, yeast, oil and water in a bowl. Knead on a floured surface until smooth. Return to the bowl, cover with a tea cloth and leave to rise for 1 hour. Stir the sauce ingredients in another bowl. Heat the oil in a pan and add the vegetables; cook on a high heat until soft.

Set oven to 240°C/475°F/gas mark 9. Divide the dough in two and roll out on a floured surface. Spread the sauce over both circles of dough and decorate with the veggies. Brush the crust and veg with oil and bake until the dough appears crispy and golden. Slice and serve garnished with rocket leaves and sliced cherry tomatoes, if desired.

Ingredients

For the base:

500 g wholewheat flour

1 tsp yeast

1½ tbsp organic olive oil

300 ml warm water

For the sauce:

120 g tomato purée

¼ tsp salt

½ tsp oregano

½ tsp dried basil

For the topping:

1 tsp organic olive oil

1 sliced courgette

1 sliced red or
   yellow pepper

750 g sliced mushrooms

1 peeled, sliced red onion

A few cherry tomatoes
   (optional)

Rocket leaves (optional)

VEGAN SHEPHERD'S PIE

Hearty, healthy and full of flavour, this plant-based alternative to the meaty English classic hits the spot on a cold day. Use puy lentils, rather than red or green lentils, to ensure the dish keeps its shape.

Serves: 2 · Preparation: 10 mins · Cooking: 25 to 30 mins

Method

Boil potatoes until tender, drain and season to taste. Crush with a masher or fork, then mix in the parsley. Preheat grill to high.

Heat the oil in a deep frying pan. Add the onion and fry for 2 to 3 minutes. Increase the heat to medium/high and add in the mushrooms, frying for 6 to 8 minutes, until soft. Stir in the lentils, tomatoes, purée, chillies and 100 ml of water. Simmer until sauce has thickened (approximately 10 minutes). Season to taste. Tip the mix into a baking dish and top with the crushed potatoes. Drizzle with oil and place under the grill until the potatoes are golden.

Ingredients

250 g peeled small potatoes

Salt and pepper

15 g chopped flat-leaf parsley

3 tbsp organic olive oil, plus extra for drizzling

½ chopped onion

300 g closed cup mushrooms

125 g ready-to-eat puy lentils

2 medium tomatoes

2 tbsp tomato purée

¼ tsp crushed chillies

GARLIC AND CHILLI MUSSELS

Simple, light and spicy, this stylish mussel dish makes a succulent starter – or serve it as a main, alongside a roast veggie salad (page 86). Ask your local fishmonger in advance for fresh mussels, farmed nearby, with the Aquaculture Stewardship Council (ASC) stamp of sustainability.

Serves: 2 · Preparation: 20 mins · Cooking: 10 mins

Method

Place the tomatoes in a Pyrex bowl and cover in boiling water. After 3 minutes, drain the water and peel. Cut the tomatoes into quarters and scoop out any seeds, then chop the soft flesh. Lightly fry the garlic and chilli in oil. Next pour in the wine and add the tomatoes, tomato paste and basil, and simmer for 3 minutes. Carefully stir in the mussels and place the lid on the saucepan. Steam for 4 to 5 minutes, gently stirring the mussels halfway through, until the shells have opened. Remove from the stove and throw away any closed shells. Serve into two bowls and decorate with any remaining basil leaves. Provide an empty bowl for shells.

Ingredients

2 medium tomatoes

1 finely chopped garlic clove

1 deseeded and chopped red chilli

2 tbsp organic olive oil

Small glass of dry white wine

1 tsp tomato paste

Handful of basil

1 kg cleaned ASC-certified mussels

BEAN BURGER WITH CARROT FRIES

Don't fall for the myth that all hearty burgers must be meat-based. This protein- and vitamin-rich vegan alternative to the classic burger and fries is the perfect go-to dinner on a lazy night in.

Serves: 2 · **Preparation**: 15 mins · **Cooking**: 30 mins

Method

Preheat the oven to 220°C/425°F/gas mark 7. Chop the carrots into chips and spread over a baking tray, drizzling with ½ tbsp of oil. Season to taste and roast (approximately 30 minutes). Meanwhile, fry the onion, garlic and chilli in 1 tbsp of oil, until soft. Pulse with the beans, herbs and old bun in a food processor (not a blender) until smooth. Add a splash or two of water to assist the pulsing if too thick. Shape and fry two 2-cm thick patties using leftover oil, until crispy. Once the carrot chips are cooked, slice two buns in half and place open in the cooling oven for 5 minutes or until lightly toasted. Build the burger: lettuce topped with a patty, cucumber and pickled cabbage – alongside any other desired planet-friendly veggies, such as red onion or home-made sundried tomatoes. Serve the carrot fries alongside.

Ingredients

250 g peeled carrots

2½ tbsp organic olive oil

Salt and pepper

½ red onion

2 peeled garlic cloves

1 deseeded red chilli

400 g drained
   kidney beans

½ tsp oregano

5 g chopped parsley

1 stale wholemeal bun

2 wholemeal buns

Handful of lettuce

¼ sliced cucumber

170 g pickled cabbage
   (see page 110)

CLASSIC CHICKEN TIKKA MASALA

Bursting with spice and flavour, this well-loved chicken curry recipe replaces some traditional ingredients with sustainable alternatives. Serve on a bed of rice – or stuffed inside a wholemeal naan bread – along with some of our spicy Bombay potatoes (page 100).

Serves: 2 · Preparation: 15 mins · Marinating time: 2 hours · Cooking: 40 mins

---

Method

To prepare the tikka, dice the chicken into a bowl and mix with the salt and lemon juice. Add the ginger, spices, oil and cream. Mix and refrigerate for at least 2 hours.

For the masala, fry the ginger and spices in oil on a medium heat in a deep frying pan for 2 minutes. Add the yoghurt and stir in until absorbed. Throw in the tomatoes and fry until pulped. Pour in the stock and simmer. Once the sauce has thickened, add the chicken and simmer further until cooked through.

Ingredients

For the tikka:

350 g organically farmed chicken

1 tsp salt

2 tbsp lemon juice

1 tbsp grated ginger

1 tsp ground cumin

1 tsp garam masala

2 tsp chilli powder

2 tbsp rapeseed oil

4 tbsp soya cream

For the masala:

1 tbsp grated ginger

1 tbsp ground coriander

2 tsp paprika

½ tsp turmeric

3 tbsp rapeseed oil

3 tbsp soya yoghurt

2 chopped tomatoes

200 ml chicken stock

SPICY BOMBAY POTATOES

Serve these delicious curried spuds either on their own, on a bed of rice, wrapped up in a warm naan bread or as a side with another Indian dish (such as the chicken tikka masala on page 98).

Serves: 2 · Preparation: 10 mins · Cooking: 35 mins

Method

Boil the potatoes in a large pan. Once tender, turn off the heat and drain well. Blend the garlic, ginger and 4 tbsp of water until puréed, add the tomatoes and blend again. Use the oil to fry the potatoes for 10 to 15 minutes until brown. Add the cumin, turmeric and mustard seeds. Leave to sizzle. Then add the puréed garlic, ginger and tomatoes and three quarters of a chilli. Cook on a medium heat for 10 minutes, until the sauce has thickened. Stir in the garam masala and add salt to taste. Serve in a bowl, topped with coriander, the remaining chilli and a lime wedge.

Ingredients

800 g peeled, chopped potatoes

4 chopped garlic cloves

Thumb-sized ginger root, chopped

4 chopped plum tomatoes

2 tbsp sustainable vegetable oil

1½ tsp cumin seeds

1 tsp ground turmeric

1½ tsp black mustard seeds

1 finely chopped red chilli

1 tsp garam masala

Salt

Handful of chopped coriander

2 lime wedges

AUBERGINE, MUSHROOM AND RED PEPPER PASTA

The fusion of red wine and these wholesome veggies creates a warming, rich and elegant evening meal. Double the garlic for a more intense taste.

Serves: 2 · Preparation: 10 mins · Cooking: 25 mins

Method

Set the pasta to boil. In another saucepan, sauté the pepper and onion in 2 tbsp of oil until soft. Remove from heat and set aside in a bowl. Add the remaining olive oil to the pan, followed by the aubergine, and cook on a high heat for 4 minutes. Add the mushrooms, chilli and garlic, and cook for a further 4 minutes, until the mushrooms wilt. Reintroduce the pepper and onion, followed by the tomato purée and red wine, and add a splash of water if needed. Stir well. Cook on a medium heat until the sauce thickens, then season to taste. Serve over the pasta, garnished with parsley.

Ingredients

250 g wholewheat/ gluten-free pasta (any)

1 chopped red pepper

1 sliced small onion

3 tbsp organic olive oil

1 diced medium aubergine

150 g mushrooms

1 deseeded red chilli

1 peeled, chopped garlic clove

3 tbsp tomato purée

250 ml red wine

Salt and pepper

1 tbsp chopped parsley

POACHED SALMON WITH MUSTARD SAUCE

This light, protein-packed salmon dish is the perfect go-to for a healthy dinner. Use sustainably caught Pacific salmon certified by the MSC – check for the label on the packaging – to ensure the meal is planet-friendly.

Serves: 2 · Preparation: 15 mins · Cooking: 20 mins

Method

Peel the potatoes and place them in a saucepan to boil until tender. Bring a shallow saucepan of water to the boil; place the salmon fillets in. Put the lid on and simmer for 5 minutes. In the meantime, blend the mustard, tarragon, parsley, vinegar and oil in a food processor until completely smooth. Add water to lighten the thickness if necessary. When the salmon is ready, carefully lift using a spatula and allow the water to drain. Serve with the potatoes, asparagus and watercress, drizzled with the mustard sauce.

Ingredients

250 g small potatoes

2 MSC-certified, skinless Pacific salmon fillets

2 tsp Dijon mustard

2 tbsp chopped tarragon

2 tbsp flat-leaf parsley

2 tbsp white wine vinegar

2 tbsp organic olive oil

10 spears asparagus (optional)

Handful of watercress

REFRIED BEANS

A side of refried beans is a healthy and satisfying addition to any meal. Packed with nutritional and planet-friendly goodness, the kidney beans can be replaced with locally grown pinto or black beans if needed.

Serves: 2 · Preparation: 10 mins · Cooking: 15 mins

Method

Fry the oil, onion, garlic and a pinch of salt on a medium heat, until the onion softens. Stir in the cumin, then add the kidney beans and water. Stir well, cover and simmer for 5 minutes. Remove from the heat and mash the beans using a potato masher. Continue cooking on a low heat, stirring often, until the mashed beans become thick and creamy. Remove from the heat once again and add the lime juice, salt and pepper, to taste. If a little dry, add a splash of water to the mix and stir. Serve as a healthy side or with huevos rancheros (page 62).

Ingredients

1 tsp organic olive oil

¼ finely chopped onion

1 peeled, sliced
  garlic clove

2 pinches of salt

1 tsp ground cumin

200 g cooked
  kidney beans

115 ml water

1 tsp lime juice

Pepper

CRISPY POTATO-PEEL CHIPS WITH SWEET BUTTERNUT SQUASH DIP

A fantastic way to combat food waste, plus the potato peelings can be swapped for sweet potato, carrot or parsnip peelings if needed. This yummy chip and dip recipe can be served as a snack or a side.

Serves: 2 · Preparation: 5 mins · Cooking: 15 mins

Method

To make the dip, add all the ingredients into a small saucepan and simmer. Once the squash is tender, pour the mix into a food processor and blend until smooth. If too thick, add a little water and blend further. If too weak, return to the heat and simmer to a sauce consistency. Store the cooled dip in the fridge for up to 10 days.

To make the chips, wash leftover peelings and pat dry with a clean tea cloth. Add 1 cm of oil to a frying pan on a medium heat and fry the peelings until crispy. (Tip: Fry in smaller batches to avoid sticking.) Drain the chips on a paper towel. Season to taste and enjoy with the squash dip.

Ingredients

For the dip:

250 g peeled, deseeded and chopped butternut squash

½ tsp coriander seeds

½ tsp cumin seeds

115 ml apple cider vinegar

100 g sustainable sugar (from beets)

Salt and pepper

50 ml water

For the chips:

Leftover potato peelings

Sunflower oil

PICKLED RED CABBAGE

A handy way to preserve veggies that are going out of date, this pickling recipe can be used for a range of vegetables, including jalapeños, carrots, radishes, peppers, beetroot, turnips and cucumbers. Try this pickled red cabbage with a vegan bean burger (page 96).

Makes: 1 jam jar · **Preparation:** 5 mins · **Cooking:** 5 mins

---

Ingredients

350 ml apple
    cider vinegar

2 tbsp organic honey

1 tbsp sea salt

1 peeled garlic clove

¼ shredded red cabbage

Method

Heat the vinegar in a saucepan until piping hot. Turn off the heat and stir in the honey, salt and garlic. Meanwhile, fill a large sterilized jar with the red cabbage. Once ready, pour in the vinegar, ensuring the cabbage is fully submerged. Place the lid on the jar while hot – this will seal the jar and preserve the cabbage. Allow to marinate for 24 hours, although the longer the better. Store in a cupboard and refrigerate after opening.

Experiment by adding other options such as expiring ginger slices, chillies or herbs to minimize waste. Peppercorns, cloves, dill or cumin seeds can also be added to suit tastes.

PARSNIP AND BEETROOT BAKE

This delightful root veggie bake with a zing of orange is a wholesome side to any meal. If parsnips and beetroot aren't your preference, simply replace them with alternative, locally grown root vegetables such as carrots and potatoes.

Serves: 2 · Preparation: 20 mins · Cooking: 1 hour 30 mins

Method

Preheat the oven to 200°C/400°F/gas mark 6. Thinly slice the parsnips and beetroots, then layer across a greased baking dish. Mix the creams, garlic and 1 stem of rosemary in a saucepan, simmering for a few minutes. Take off the heat, add the zest and season with salt and pepper to taste. Pour into the baking dish, submerging the veggies. Arrange the remaining sprigs of rosemary on top. Cover the dish tightly with kitchen foil and bake. After 45 minutes, remove the foil, before returning to the oven for a further 30 minutes. The bake is ready once the top layer is set and there is browning at the edges. Allow to cool for 10 minutes before serving.

Ingredients

3 large peeled parsnips

2 scrubbed beetroots

150 ml double soya cream

100 ml single soya cream

2 whole garlic cloves

3 stems rosemary

Zest of 1 orange, grated

2 tsp sea salt

Black pepper

SWEET POTATO HASH

This delicious and nutritious vitamin-loaded hash can be enjoyed at both breakfast – serve with poached or fried organic eggs – and dinner, alongside a vegan source of protein or sustainably sourced fish.

Serves: 2 · Preparation: 5 mins · Cooking: 25 mins

Method

Chop the peeled sweet potatoes into small chunks, add to a small pan with water and bring to the boil. Allow to simmer for 5 to 7 minutes, until they start to soften. Be careful not to overboil them. Remove from the heat and drain. Using a frying pan, heat the oil and fry the onion and garlic until golden. Add in the pepper and sweet potatoes. Cook on a medium heat for approximately 10 minutes, or until the sweet potatoes start to brown. Once ready, serve on a plate or in a bowl, seasoned with salt and pepper to taste and a sprinkle of chopped parsley.

Ingredients

2 large peeled sweet potatoes

2 tbsp rapeseed oil

1 large peeled and diced onion

2 peeled garlic cloves

1 large red/yellow pepper, deseeded and chopped

Salt and pepper

Small handful of parsley

CRISPY KALE CHIPS

It's simple to make, 100 per cent healthy, totally moreish – and a completely planet-friendly snack. These crispy kale chips are perfect for a pre-meal snack or for those who love to constantly graze.

Preparation: 5 mins · Cooking: 15 mins

Method

Preheat the oven to 180°C/350°F/gas mark 4. Line a baking tray with baking paper and distribute the kale evenly. Lightly drizzle with olive oil and sprinkle with salt and cumin. Bake for 10 to 15 minutes, or until the kale darkens and becomes crispy. Allow to cool before serving in a bowl.

Ingredients

1 bunch of kale, cut into large pieces

1 tbsp organic olive oil

½ tsp salt

2 tsp cumin

FRUITY RICE PUDDING

Dairy-free, gluten-free and free of refined sugar, this gloriously creamy rice pudding shows that even the most delectable desserts can still be healthy and planet-friendly. Delicious when served hot or cold.

Serves: 2 · Preparation: 5 mins · Cooking: 50 mins

Method

Add the rice to a pan and cover with water. Once boiling, turn the heat to low and allow the water to reduce for 10 minutes, until the rice is soft. When ready, drain any remaining water. Add the coconut milk and cook on a medium to low heat for 20 minutes, stirring frequently to stop the rice from sticking. Add the honey and cook on low for a further 10 minutes, stirring until the liquid is mostly evaporated. Present a ladleful of pudding in a bowl and decorate with chopped strawberries or your preferred fruit. Sprinkle with cinnamon and serve.

Ingredients

60 g jasmine rice, rinsed and drained

500 ml coconut milk

1 tbsp organic honey

4 strawberries (or any other seasonal fruit)

Pinch of ground cinnamon

CINNAMON AND CHILLI PEARS

This divine dessert is a mouth-watering finish to any meal. Replace the pear with seasonal and locally grown fruits such as apple, peach, apricot or nectarine if desired.

Serves: 2 · Preparation: 5 mins · Cooking: 10 mins

Method

Peel and core the pears, then chop into halves. Cook in a frying pan over a medium to high heat, turning every 2 minutes, until they begin to soften and yellow. Ensure the pears do not burn by adding a splash of water if they start to dry out. Once naturally caramelized, reduce the heat, add the honey, cinnamon and chilli, if desired, and stir gently with a wooden spoon. Allow the mixture to bubble and check the spiced honey dressing regularly to ensure the consistency is a sticky, medium thickness. If overly light, keep heating until thickened. If too thick, add a splash of water. Once ready, serve the pears on a small plate with a dollop of yoghurt.

Ingredients

4 pears

1 tbsp organic honey

½ tsp ground cinnamon

½ tsp chilli flakes (optional)

Soya or coconut yoghurt

SWEET ORANGE PANCAKES

You can't go wrong with a serving of warm orange pancakes, rolled up and oozing with citrus flavour. The ideal sweet treat for a weekend or post-dinner pick-me-up – and full of vitamin C too!

Serves: 2 · Preparation: 10 mins · Cooking: 10 mins

Method

Sift the flour into a large bowl and make a small indentation in the centre. Add the eggs and whisk, while gradually adding in the milk. Once the batter is ready, leave to stand for 10 minutes. Grate the orange zest into a bowl then remove the remaining skin. Segment the oranges then add to a small saucepan along with the zest. Cook until the oranges are softened, adding a splash of water to retain moisture. Remove from the stove. Pour the batter into a non-stick frying pan, swirling to achieve an even spread. Cook for 1 to 2 minutes on each side over a medium heat, until golden. Once cooked, place the pancakes on a plate and serve with four orange segments and the zest. Drizzle with honey.

Ingredients

200 g wholewheat flour

2 organic eggs

300 ml unsweetened plant milk

2 oranges

1 tsp organic honey

BLACKBERRY AND APPLE CRUMBLE

Warming, sustainable, delicious and packed with the good stuff. You can swap the fruits in this gloriously satisfying crumble for similar seasonal and local produce to keep it super planet-friendly.

Serves: 2 · Preparation: 5 mins · Cooking: 40 mins

Method

Place the oats in a large bowl. Heat the coconut oil, honey and cinnamon in a pan until syrupy then stir into the oats to create a sticky crumble mixture. Leave to rest. Chop the apples into small bite-size chunks and add to another saucepan together with the blackberries and water of approximately 1 cm depth. Add the honey and cinnamon and leave to simmer with the lid on, until the apples are soft. Meanwhile, preheat the oven to 180°C/350°F/gas mark 4. Once the fruit is ready, transfer it to a small baking dish and level off. Top with a generous layer of oat mix. Bake for 30 minutes, until the top is a crispy brown.

Ingredients

For the crumble:

60 g oats

1 tsp sustainable coconut oil

30 ml organic honey

1 tsp ground cinnamon

For the fruit filling:

2 peeled red apples

150 g blackberries

½ tbsp organic honey

½ tsp ground cinnamon

CONCLUSION

Thank you for reading *The Planet-Friendly Kitchen*. Hopefully, by shedding light on how our food is grown and made, and the impact those processes have on the planet, it's inspired you to make a change toward a more planet-friendly lifestyle.

As consumers, we have a great deal of power to shape the future of our food industry – and this shift will be vital as the global population grows. With our food being so intertwined with the welfare of our planet and its ecosystems, even small changes to our diet can make an almighty difference. By supporting and shopping for produce from the farms and distributors who follow sustainable, organic and ethical practices, we can help cultivate a kinder, more planet-friendly way of feeding the world.

Perhaps by using the information in this book, you can enjoy an increasingly planet-friendly diet, bearing in mind that being "planet-friendly" is not only about the food we buy – but also about how we store, cook and consume our food. You may feel empowered to cultivate your own vegetable and herb garden or take a foraging course in your local area. You might even consider switching to a predominantly plant-based diet, shopping locally for your groceries and meal-planning to reduce food waste.

We know the planet's resources are already stretched to their limits. Without significant changes to how we produce and consume food, our wildlife and the environment will continue to suffer. Making good choices now will help to preserve this beautiful world for our children, grandchildren and many generations to come. By opting to be a part of this change, you are doing something truly wonderful!

If you're interested in finding out more about our books,
find us on Facebook at **Summersdale Publishers**
and follow us on Twitter at @**Summersdale**.

www.summersdale.com